HMS Victory.
Photo: University of Portsmouth

Southsea.
Photo: David Burtenshaw

Spitbank Fort.
Photo: David Burtenshaw

Eastney.
Photo: David Burtenshaw

St Michael's Building, University of Portsmouth.
Photo: University of Portsmouth

Old Portsmouth.
Photo: David Burtenshaw

Gunwharf Quays.
Photo: David Burtenshaw

Portsea Island.
Photo: University of Portsmouth

Discovering Cities

Portsmouth

David Burtenshaw
University of Portsmouth

Series Editors
Peter S. Fox and
Christopher M. Law

Geographical Association

Preface

© University of Portsmouth

The variety and complexity of cities as revealed in their built form has been a source of fascination to the local resident and visitor alike. Can a clear spatial structure be discerned? Why do activities cluster in distinctive quarters or zones? How do relict features throw light on the constantly evolving city?

For a long time, human geographers, regional economists, urban sociologists and local historians have sought to understand the processes which shape the city. The growth (or decline) of the city is affected by local, regional and global economic forces. The forces which shape the internal structure of the city are many and varied. There is a market in land which influences the pattern of land use and change. Public policies are often significant but can be complex to understand and difficult to follow. Social factors such as those of class and ethnic community identities are also important.

Written by urban geographers with vast knowledge and experience of the city in question, *Discovering Cities* gathers these issues together in concise and practical guides, illustrated with colour maps and photographs, to enable an enhanced perspective of cities of the British Isles.

Peter S. Fox, Chilwell Comprehensive School, Chilwell, Nottingham

Christopher M. Law, Visiting Fellow, University of Salford and Research Associate, University of Gloucestershire

Acknowledgements
Dedicated to the memory of Tudor Rawkins and Colin Warner, founders of the Geography Department in Portsmouth.

ISBN 1 84377 057 1
First published 2002
Impression number 10 9 8 7 6 5 4 3 2 1
Year 2004 2003 2002

Published by the Geographical Association, 160 Solly Street, Sheffield S1 4BF.
Website: www.geography.org.uk
E-mail: ga@geography.org.uk

The Geographical Association is a registered charity: no 313129.

The Publications Officer of the GA would be happy to hear from other potential authors who have ideas for geography books. You may contact the Officer via the GA at the address above. The views expressed in this publication are those of the author and do not necessarily represent those of the Geographical Association.

Editing: Rose Pipes
Design and typesetting: Arkima, Leeds
Cartography: Paul Coles
Printing and binding: Stanley Press, Dewsbury

© David Burtenshaw

Contents

Introduction 8

Site and early growth

Historic origins 10
The nineteenth century 12
The twentieth century 15

Economic and social geography

Post-war recovery 18
The city at the Millennium 21
Marshes and morasses 23
From militarism to consumerism 26
Tourism: from resort to defence heritage 31

Small area studies and trails

Introduction 37
Trail 1: Heritage trail 37
Trail 2: Southsea trail 42
Trail 3: Tipner 45
Trail 4: Gosport heritage trail 48

Bibliography and further information 53

Introduction

Figure 1: Portsmouth site and situation.

The key to discovering Portsmouth is its unique physical setting (Figure 1) and how that has been used in the defence of the country. Portsmouth has long been the home of the Royal Navy and the city has evolved over the centuries in line with the defensive needs of the country. The site on Portsmouth Harbour provided the environment in which the city grew, where defensive priorities altered landscapes and shaped the modern city, one of the most densely populated in Europe. Today, it is the re-utilisation of the landscapes of military might that is a core theme in discovering the city. It is an island city dominated by past townscapes until the twentieth century when Portsmouth was 'off the island'. Only now, at the turn of the century, has the city rediscovered and reinvented itself.

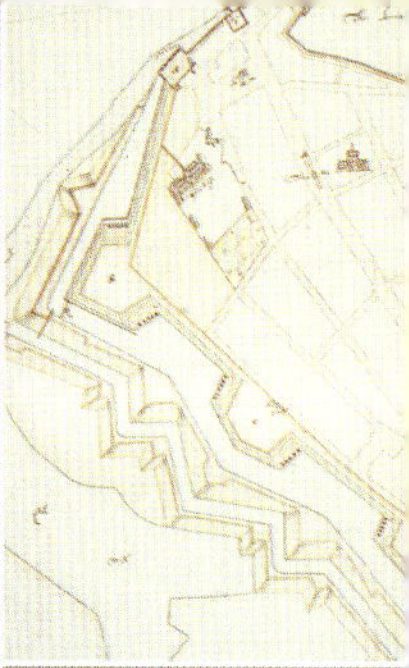

© By permission of the British Library

Site and early growth

Contents

vid Burtenshaw

Historic origins 10

The nineteenth century 12

The twentieth century 15

© University of Portsmouth

© English Heritage Photo Library

Figure 2: Portsea Island.

Historic origins

Portsea Island is low-lying, rarely rising more than 5m above sea level (Figure 2). It is surrounded by the drowned river channels of streams that drained the chalk to the north and flowed into the main Solent River channel along the course of the present-day Solent, separating the Isle of Wight from the mainland. The south coast of Portsea Island was a line of shingle barrier beaches fronting marshland. To the west and east the coastlines were indented with small, marshy, partly sediment-infilled valleys whose streams drained into either Portsmouth or Langstone Harbour. The courses of these streams and the remnants of the marshlands are important elements in the evolution of the cityscape. At the mouths of the harbours there are classic recurved spits that have been stabilised by development, each providing a unique element in the geography of the city. The northern shore of the island, Ports Creek, is now barely 200m wide but it remains a significant divide in the landscape.

The backdrop to the city is Portsdown Hill, a 130m-high, chalk pericline extending 12km from west to east (Figure 3). This is a further key landscape element in the city's evolution. People in the city still refer to those living north of Portsdown Hill as being 'over the hill', an image of a detached area.

West of Portsmouth Harbour lies the Gosport peninsula whose harbour

© David Burtenshaw

Figure 3: Portsdown Hill.

© David Burtenshaw

frontage is still characterised by creeks stretching inland and dividing up the peninsula. Much of the southern half of the peninsula is underlain by Plateau Gravels. Across Langstone Harbour to the east lies Hayling Island, physically similar to Portsea, and beyond that the reaches of a third major harbour, Chichester Harbour.

Portsmouth did not exist as a town before the twelfth century although the Romans fortified Portchester to the north of Portsmouth Harbour. A small settlement grew beside the Camber, an enclosed safe haven protected by recurved spits at the entrance to Portsmouth Harbour. The town charter was granted by Richard I in 1194. Portchester Castle became a royal settlement and Portsmouth Harbour the port from which kings set sail for expeditions to Normandy, Poitou and Gascony. As the upper reaches of the harbour silted up so Portsmouth's role was enhanced. A consequence of the town's key role in relation to France was frequent raids by the French, necessitating fortification. The oldest fortification, The Round Tower dating from 1418, is still a prominent feature. One notable raid resulted in the sinking of the Mary Rose (1545) at Spithead in full view of Henry VIII in Southsea Castle. The vessel was recovered in 1982 and is one of the city's many maritime heritage attractions.

The dockyard had existed since the time of King John. It owes its growth to the decision by Henry VIII to build the first dry dock in the world. Timber for

Figure 4: *Plan of the fortifications of the town of Portsmouth, Sir Bernard de Gomme, 1677*

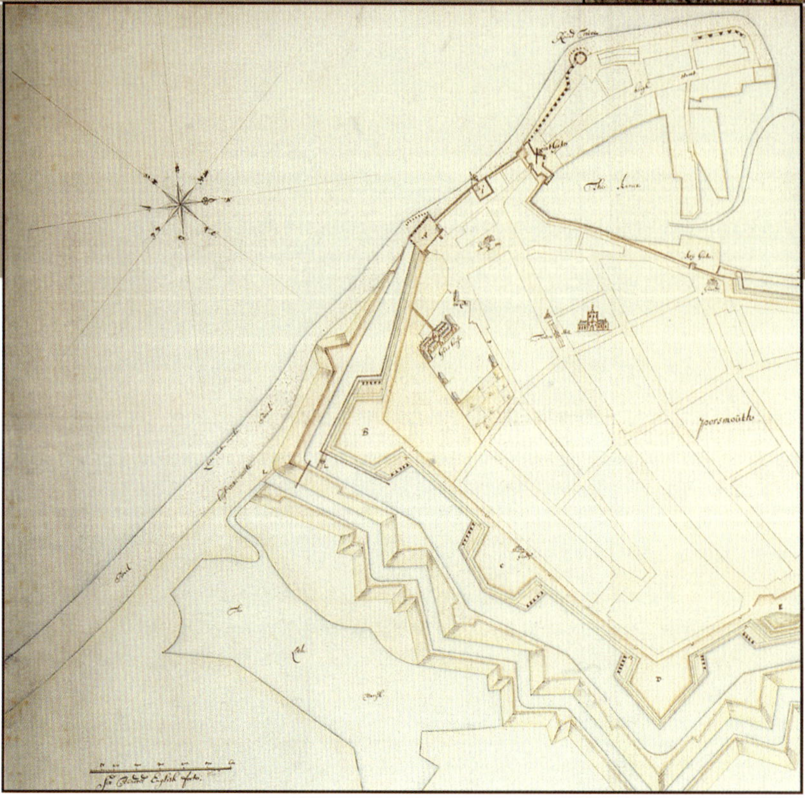

warships was readily available in the Forest of Bere, north of Portsdown Hill, and in the Weald to the east. In the seventeenth century the town became highly fortified in response to improvements in naval artillery (Figure 4). On the seaward side the Square Tower, ramparts and a massive wall gave protection. Inland a series of complex earthworks were constructed by de Gomme, a Dutch engineer, who also fortified Gosport. The Point, the spit protecting the Camber, was outside the walls and almost uninhabited until the seventeenth century. Just to the north of the town was an inlet, the Mill Pond, that separated the town from the early dockyard. To the east the land was marshy and close to sea level except for the beach bars on which Southsea Castle (1544) had been constructed (Figure 5).

Workers were attracted to the dockyard and a new settlement north of the Mill Pond and adjacent to the Dockyard, Portsea, grew to house them. It was fortified and had outgrown its neighbour, Portsmouth, by 1801 with a population of 24,327 compared to 7839 in Portsmouth. Already the dockyard worker's settlement had grown distinct from the commercial town.

The nineteenth century

By 1801 further fortifications had been developed on the southeast of Portsea Island at Fort Cumberland, and along the northern edge of the island, Hilsea Lines. Napoleonic prisoners of war were used as labour to drain the common and marshland of Southsea and roads were laid out across it. The drained land was

© David Burtenshaw

Figure 5: Southsea Castle.

© David Burtenshaw

owned by the War Department until 1923 when it was sold to the city to become modern Southsea Common. After 1809 the first areas of Southsea were developed – Landport, Hampshire, Kings and Bellevue Terraces overlooking the old ramparts. Later, substantial villas to house the officer classes were built, notably by Thomas Ellis Owen between 1835 and 1860. By 1830, 50% of the 'nobility' were recorded as moving to Southsea. Fashionable Southsea had become the third element in the social geography of the city.

Throughout the nineteenth century the city grew in tandem with the defensive needs of the Empire. A canal had been dug across the island to link the harbour to London via the Chichester and Wey and Arun canals avoiding passage through the English Channel. However, by 1859, the arrival of the direct railway from London (the railway from Brighton reached the city in 1847) made the canal redundant. The railway used part of the former canal for its route between Fratton and the town station. The railway assisted the development of the resort function of Southsea. Piers were constructed in 1860 (Clarence Pier) and 1879 (South Parade Pier). A short-lived (1885-1914) branch line led from Fratton towards Southsea and its course can still be traced in the modern street plan.

In the 1850s and 1860s, the continuing fear of the French led to a ring of defensive fortifications surrounding the whole island. Four sea forts were erected. The earlier fortifications around Old Portsmouth and Portsea, made redundant by new gunnery techniques, were demolished by 1878. The cleared land was used for parks, barracks and military recreation grounds. A further line of six forts were built along Portsdown Hill to defend against attack from inland. The Gosport peninsula was

Table 1: Population data 1801-2000 (est).

	Portsmouth	Gosport	Havant	Fareham	Hayling
1801	32,166	11,295	1670		578
1811	40,567	12,212			620
1821	45,648	10,972			738
1831		12,367			882
1841	53,032	13,512	2101	34,953	
1851	72,096		2416	36,296	
1861	94,799	22,653	2470	45,001	1039
1871	113,569	22,637	2634		1139
1881	127,989	21,581	3032	7183	1334
1891	159,251	25,452	3474	7134	1495
1901	190,281	28,879	3637	8246	
1911	233,573	33,300		11,778	
1921	247,284	33,580		10,063	
1931	249,283			11,595	
1951	233,543	60,000	35,416		
1961	215,077		74,552		
1971	196,950		108,320		
1981	175,382	76,757	116,101		
1991	169,834	73,241	117,342	100,495	
2000 (est)	192,100	85,000	120,335	107,746	17,431

N.B. In all cases boundary changes have enlarged the administrative area of each place.

similarly defended by five forts along the Elson-Gomer line. Forts on the Isle of Wight and in the western Solent (Hurst Castle) completed the nineteenth-century defensive shield. By the time they were all completed, improved Anglo-French relations and changed military technologies rendered them redundant. They became aptly named 'Palmerston's follies'. All of these elements are significant in the landscape today because 60% of the island was owned by the military. It is the return of land to civilian uses that is a key theme in the area's geography since 1945.

The dockyard covered 121.5ha in 1876 including the excavated new docks and reclaimed land. The excavations resulted in the creation of Whale Island from dumped marine mud. Each phase of development demanded labour which migrated to the city. By 1871 the population was 113,569 rising to 159,251 20 years later (Table 1). Dockyard workers (15,000 in 1914 rising to 23,000 in 1917) formed half of the industrial population of the city while service personnel numbered 17,567. Industry and the Dockyard both employed 27% of the workforce in 1911. Immigrants were housed in terraced housing which stretched across the island towards Fratton, Milton and north to Stamshaw, North End and, later, Copnor. Artisan Southsea (Somerstown) joined these new areas to the resort. All of these developments have been aptly described by Patricia Haskell as 'Portsmouth's cramped monotony of straight rows of red-brick houses tailored to the occupants' income'. Little development spread east of the railway line because there were few crossing points. Only hospitals and 'the lunatic asylum' were found to the east. By 1910 the pattern of military land ownership was declining to 30% of the island. This remained the case until after 1945. Clothing manufacture for the military who had to buy their own uniforms, and corset manufacturing in particular, were the main sources of work for women, especially for widows.

© Portsmouth City Council

© University of Portsmouth

Southsea terraces.

Brewing was the sole other industry of importance.

The twentieth century

After 1918 shipbuilding declined to be replaced by rehabilitation and refitting of warships. Dockyard employment rose again as rearmament took hold before the Second World War but the numbers employed were not to reach their former peak. Housing for the growing population spread 'off the island' to Cosham and, with the stimulus of a tramway, in a ribbon along the A3 towards Horndean. Similar developments were attached to Gosport, Fareham and Havant. A small commercial airport was opened in 1932 on 275ha to the northeast of the island with access provided by the second crossing (Eastern Road) onto the island, opened in 1941.

The Second World War brought much physical damage as a result of air raids which destroyed 6,625 homes and damaged a further 6,000. There was little space to house both the displaced population and the returning servicemen. Seven thousand homes were scheduled for slum clearance. In 1944 Portsmouth bought 676.7ha at Leigh Park beyond the city for one of the UK's largest overspill council estates. A further estate was developed at Paulsgrove. By 1980 40,000 people lived in overspill estates at Leigh Park, Crookhorn and Wecock Farm. Private housing developed along the main routes so that the once identifiable towns of Fareham, Gosport, Havant and Waterlooville became part of a coastal conurbation. Dormitory suburbs such as Stubbington, West Fareham, and Cowplain added to the sprawl while some of the nearby villages such as Clanfield and Rowlands Castle became classic commuter villages. Portsmouth is underbounded – the official city boundaries do not encompass the whole built-up area. The construction of the A3(M) and M27/M275 aided commuting. Viewed from Portsdown Hill the land to the north-west is rural, whereas east of the A3 axis it is predominantly urban. The reason for this is partly that no major routes cross the western hill. The landowners to the west, and particularly the owners of the Southwick estate, chose not to sell at the time when the landowners of Leigh Park, Purbrook and Waterlooville elected to sell. Today this basic west-east distinction in the landscape north of Portsdown Hill is embodied in the Hampshire Plan.

© Explosion

Gunwharf.

© David Burtenshaw

In the latter half of the twentieth century the engine of growth and change was no longer defence, although its presence remained considerable. There was a need to attract new employment after 1945. The first industrial estate was opened at Fratton and others followed at the airport, Farlington, Paulsgrove and Leigh Park. Electronic engineering was set to become the major employer in the region, often supplying military hardware and weapons systems. The development of IBM's manufacturing plant at Havant in 1964 was followed by the relocation of its headquarters from Chiswick to Cosham and North Harbour in the 1970s. Redundant military lands became the focus for developers for housing (Hilsea), for the new commercial and cross-Channel port (1976) at Mile End and the development of the University. The release of military-owned land, including the recent release of lands around Portsmouth Harbour, has encouraged the development of museums (Priddy's Hard), housing and retailing (HMS Vernon site, now called Gunwharf Quays) and marinas (Gosport Marina). By 1986 military land ownership had declined to 10% of Portsea Island.

The airport closed in 1967 and its release provided further land for industry (Alenia-BAE Systems) and housing (Anchorage Park, 1981). The release of nineteenth century health facilities has provided further brownfield sites. Former isolation and mental hospital facilities at Milton are undergoing redevelopment for a secondary school in the former case and awaiting development in the latter case. Fratton goods yards await redevelopment and the possible relocation of the football stadium. Remote from the city and its nearest town, Fareham, the former mental hospital at Knowle is being redeveloped as a 'new village'.

The themes that one can explore in Portsmouth are:

1. The impact of physical geography, in the form of reclamation of tidal inlets, on modern urban morphology;
2. The continuing impact of past wetlands on present day urban issues;
3. The reuse of redundant military land; and
4. The development of the heritage industry from the relics of past military glories, replacing the seaside resort function.

© Portsmouth city Council

Economic and social geography

© University of Portsmouth

Contents

Post-war recovery 18

The city at the Millennium 21

Marshes and morasses 23

From militarism to consumerism 26

Tourism: from resort to defence heritage 31

© Portsmouth City Council

Figure 6: Portsmouth wards, 1991 and changes, 2002.

To understand the present social geography of the city the key shifts in the post-1945 economic evolution of the city need to be explained. The changing economic fortunes of the city are reflected in the range of developments where the combined effects of global, international, national and local forces have shaped modern Portsmouth and its region.

Post-war recovery

In 1945 Portsmouth was a city in ruins yet a city tasting success as it went on to be champion of the Football League in 1948-9. The city contained swathes of unsatisfactory dwellings and bomb damaged areas. It was these areas which became the sites for early comprehensive redevelopment in the 1960s, e.g. Somerstown and Landport.

© University of Portsmouth

Figure 7: Deprivation map of Portsmouth, 2000.

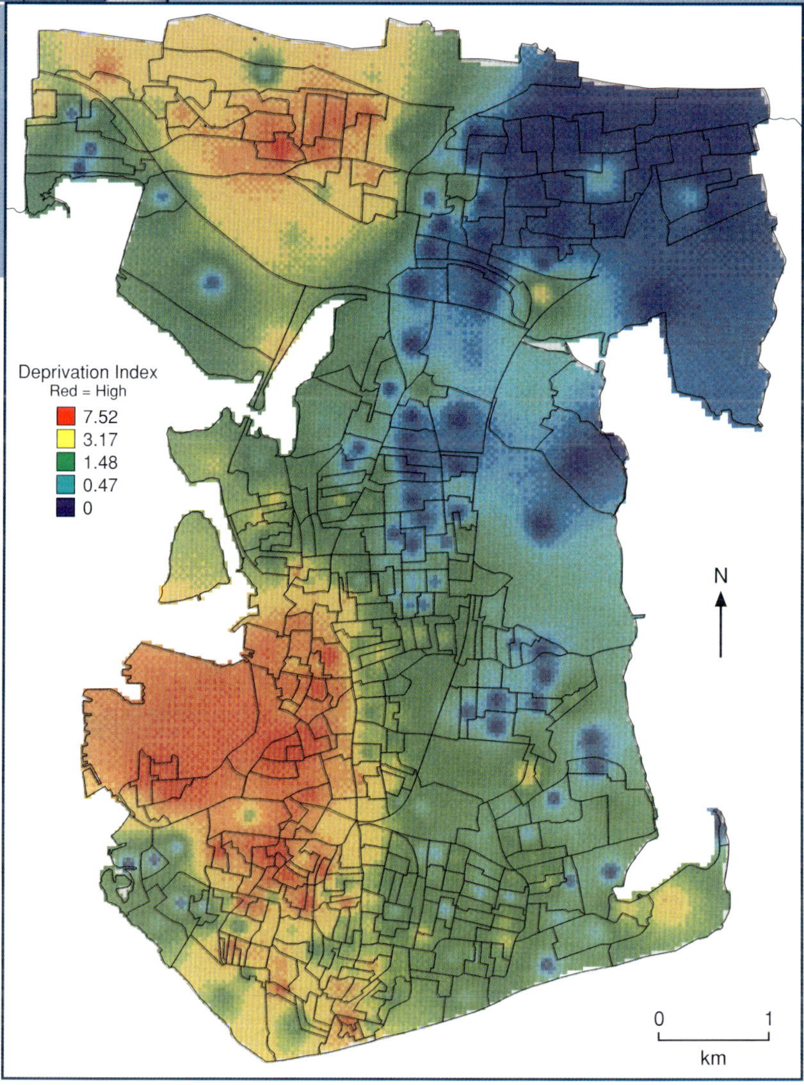

By the 1950s the realities of the atomic age, the impending decline of the Empire and financial constraints made it plain that the city could no longer depend on the military. A city with very few other sources of employment needed to develop alternatives. Initially, these were located in new industrial estates and later on land around the airport, Paulsgrove, Cosham and Farlington, where a range of electrical engineering and manufacturing firms were located. Other industrial estates developed beyond the city in Leigh Park, Waterlooville and Fareham.

Decentralisation

A much larger step was taken in the 1960s when global companies moved into the region. IBM opened a major plant in Havant and Plessey, GEC and Marconi all reinforced the dominance of

Figure 8: The socio-political geography of Portsmouth.

	Paulsgrove	Cosham	Drayton & Farlington	Nelson	Hilsea	Copr
Population	15,679 (26% 0-15yr old)	12,652 (20% 0-15yr old)	12,683 (23% retired)	15,609 (22% 0-15 yr old)	16,505	15,4
Deprivation	2nd most deprived	-	Least deprived	3rd most deprived	-	2nd depr
Density (persons per hectare)	31.7	25.8	22	67	52	36.6
Housing characteristics	Post 1945, Port Solent post	East Paulsgrove council, Highbury & Cosham 1930s	Interwar larger	Stanshaw pre 1910, Buckland redevelopment post 1970	Interwar with some post 1970	Mair C20
% Unfit dwellings	1980	6	1.7	28	6.5	8
Private housing	25	78	91	75	82	99
% open space	58 19 (Portsdown)	18 (Portsdown SSSI)	32 (Farlington Marshes)	7	17 (Hilsea Lines, Alexandra Park)	29 (and zone Lang harb
Average gross income	£16,000	£19,600	£23,000	£16,100	£19,200	£21
% Unemployment (2000)	3.6 (3.3% HE qual.)	2.2	1.2	3.3 (7% HE qual.)	2.1	1.5
% Jobs in city in ward	4	14	7	4	6	11
Significant employers	Vosper, Torneycroft	IBM, QA hospital	Industrial estate	Continental Ferry Port, Navy	Industrial estates	BAE
Councillors (2002)	3 Labour	3 Labour	3 Conservative	3 Labour	2 Conservative, 1 Labour	3 Cc

electronic engineering. More significantly, Portsmouth became involved in the 1960s decentralisation of offices from London. The initial move was by an engineering company but the significant moves were those of Zurich Insurance to Portsea (now relocated to the city centre) and IBM to Cosham prior to acquiring the reclaimed polder at North Harbour for their UK HQ. Today, Portsmouth has the European HQ of PALL, and major factories of GEC and Alenia-BAE systems.

Educational excellence

The 1970s also saw the first stages in the development of the city as an educational centre with the establishment of the polytechnic, which became a university in 1993. Today, 17,000 students have given many parts of Southsea the feel of an urban student village. Halls of residence cluster in the city centre and at Langstone and much retailing has become student-centred, e.g. the bars, take-aways and small shops of Albert Road. Student Southsea has partly taken over the northern flanks of traditional tourist Southsea.

The continental ferry port

The other major development was the opening of the Ferry Port at Mile End in 1976. Ro-Ro routes to Le Havre, Cherbourg, St Malo, Caen, Bilbao and

© David Burtenshaw

	Fratton	Milton	Highland	Havelock	St Jude	St Thomas	Charles Dickens
	14864 (High 16-44yr old)	13,548 (22% retired)	15,104	13,307	12831 (26% retired)	16867 (24% retired)	16287 (22% retired)
			Low	Pockets	-	Significant areas of deprivation	Most deprived. Acute child poverty, 9.9% single parent households
	112	40.4	58	128	73.4	84	50
	Pre-1910 in need of revival	Early C20th Mod	Mixed	Student Southsea' C19th terraces	Student Southsea, Owen's Southsea	Old Portsmouth & Somerstown council flats	70% council
	10	6.7	9	11	14	3.5	5
	90	87	89	96	98	73	25
se	20 (cemeteries and brownfield)	15 (Langstone coast zone)	18 (sea front)	<1	23 (Southsea Common)	8	6
	£19,000	£19,000	£21,800	£18,900	£22,000	£20,500	£12,000
	2.9	2.1	2.6	3.6	5.7 (18% HE qual.)	5.7 (18% HE qual.)	2.1
	1	10	<1	<1	4	8	32
	No large employers	University campus, hospital	No large employers	No large employers	Hotels	University campus	University campus, City Centre, Gunwharf, Navy
ve	3 Lib-Dem	3 Lib-Dem	3 Lib-Dem (Eastney & Craneswater)	2 Labour, 1 Conservative (Central Southsea)	3 Conservative	3 Conservative	2 Labour, 1 Lib-Dem

the Channel Islands brought a steady stream of passengers and freight through the city. In 2000, 3.3 million passengers and almost 1 million vehicles passed through the port.

The city at the Millennium

At the turn of the century Portsmouth is no longer a defence city. It is reinventing itself as a service-oriented city, dependent on education, tourism and high technology employing skilled people. Its population is still about 190,000 but there are now 630,000 people living in the surrounding districts and the Isle of Wight.

The modern social geography of the city is much influenced by the past and the basic patterns laid down in the nineteenth century remain: Portsea is a lower socio-economic enclave; Old Portsmouth has gentrified and is no longer the commercial heart of the city; Southsea, a faded resort whose core remains a middle-class enclave, now has its own town council. The university population has reinvigorated the northeastern fringes. Over 20,000 properties from Eastney to Fratton and north to Stamshaw and Tipner have benefited from Housing Action Area and General Improvement Area improvements and are now in need of 'housing revival'. Some of these areas,

Figure 9: Marsh reclamation.

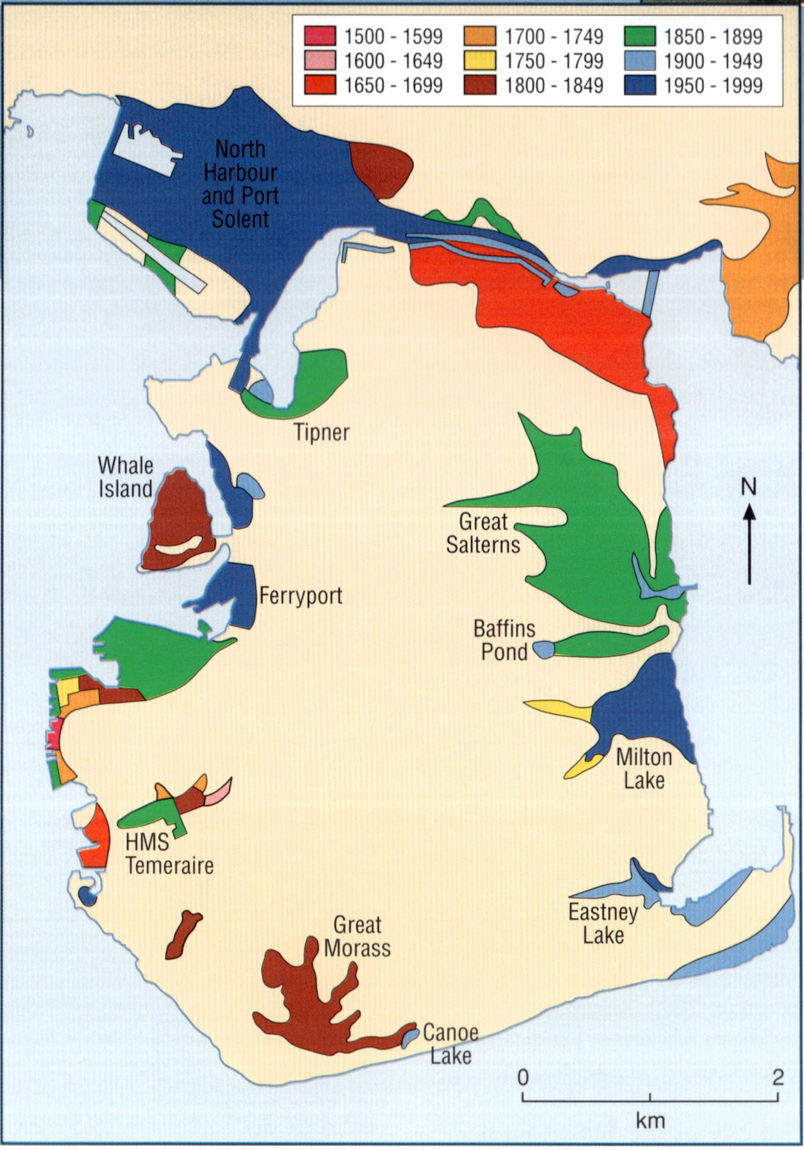

e.g. the former Havelock ward (Figure 6), are increasingly the domain of professional single people.

The city remains an anomaly in affluent South East England. It is a relatively deprived island surrounded by the affluence of Hampshire and West Sussex. The map of deprivation in 2000, which combines income, employment, health, education, housing and access to services, shows the basic divide between west and east (Figure 7). Within the west there are some enclaves of affluence in Old Portsmouth, Port Solent and Portchester. In contrast, there are few enclaves of deprivation in the affluent east with the exception of parts of Eastney. Two wards, Charles Dickens and Paulsgrove, are ranked in the worst 1000

© University of Portsmouth

Figure 10: Eastney.

© David Burtenshaw

nationally, at 601 and 956 respectively. On the dimensions of deprivation such as education and skills, eight wards are in the UK's worst 2000. Six wards are in the worst 2000 for housing. Most alarming is the fact that Charles Dickens is the 23rd worst ward in the whole country for child poverty, with 80% of the children being from families with means-tested benefits. The main characteristics of the city's socio-economic geography in 2000 by ward are shown in Figure 8.

Marshes and morasses

An eighteenth-century writer noted that Portsmouth seemed 'to be in the very water'. Today, the former marine inlets surrounding Portsea Island are a significant feature in the city's modern form. The reclamation of many of these areas (Figure 9) over the past two hundred years has provided opportunities for development and open space, and has led to problems of pollution and flooding.

Most of the southwest fringes of the island were reclaimed for the dockyard with the notable expansion of Whale Island from the mud excavated for the dry docks. The reclamations at Tipner (1900), the Ferry Port (1971) and Gunwharf (1690) are referred to later. To the east the largest reclamation was that of Great Salterns, formerly used as saltpans for supplying the fleet. Place and street names around the city point to the profusion of minor watercourses whose significance is lost until disaster strikes; e.g. Lake Road, Hambrook Street and Craneswater. Few vestiges of these water courses and marshlands remain. The most significant are Canoe Lake, the surviving remnant of the Craneswater (Crana Lake) arm of the Great Morass and Baffins Pond, a remnant of Gatcombe Haven.

Discovering reclamation

1. The HMS Temeraire fleet recreation ground is sited on part of the Mill Pond that was successively reclaimed between 1650 and 1860. The sole vestige of the inlet is the small dock inlet now a water feature in the Gunwharf Quays development.

Figure 11: SSSI and Ramsar sites.

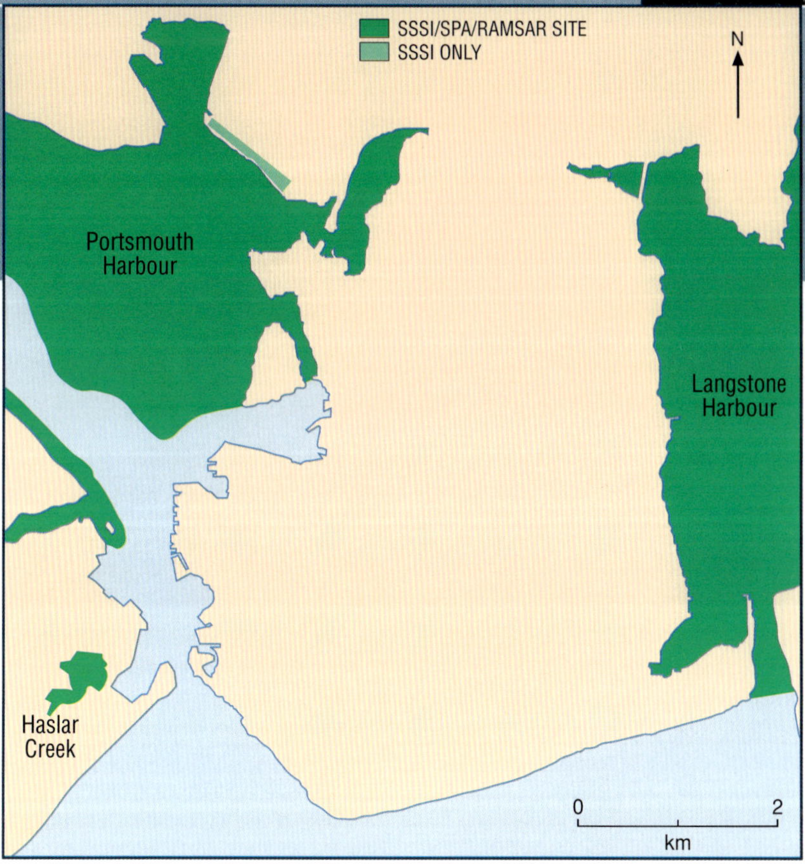

2. The Great Morass (see Trail 2) was drained between 1823 and 1845 and some of the western fringes were built upon by 1870. By 1910 much of the area was developed with the lowest lying areas remaining as open space (Clarendon Gardens and Waverley Gardens) or in transport uses such as a garage and the former Southsea railway station. Since 1960 these sites have been redeveloped for housing and some open space has been used for St Swithuns School.
3. Approximately 6.7ha of Eastney Lake were reclaimed by dumping waste from the dockyard during the interwar period, so much so that it acquired the nickname 'Glory Hole'! Ministry of Defence (MoD) housing was built on part of the site in the 1960s. Soil samples taken later revealed that toxic wastes were dumped on the site although the military records were scant. Protective membranes have been laid beneath the topsoil to prevent the further capillary movement of pollutants. It cost £2.25 million to remove asbestos and heavy metals from the site of Langstone Marina. Some private housing has been built on the site (Figure 10).
4. The inlet which formed Milton Lake was reclaimed between 1962 and 1970 by the dumping of the city's waste. The land has been left as

© Portsmouth City Council

Port Solent.

© University of Portsmouth

open space while the rubbish beneath decomposes. Methane extractors can be seen alongside Eastern Road. All of Langstone Harbour to the east is a site of special scientific interest (SSSI) and a Ramsar (wetland) site (Figure 11).

5. By far the largest scheme has been the reclamation of 182 hectares at North Harbour and the adjacent Port Solent Marina. The stimulus for this development was the creation of the M27/M275 junction and the M275 route into the city. Modern polder reclamation techniques were used to create the embankments for the road using sands dredged offshore and carried into the harbour. North of the motorway is the 51ha site of IBM's UK Headquarters. This was a classic case of 1970s office decentralisation from London. A lake associated with the balancing of the water table on the site has become a nature reserve, a symbol of greater environmental awareness. South of the motorway to Horsey Island, itself reclaimed from two islands to create a torpedo range, lies Port Solent. This was developed by Arlington Securities, a major property development group. Here a mixed-use development has been constructed focusing on the marina. Offices, housing, retailing, waterfront cafes and a multiplex cinema have been built together with associated car parking for this site which is not very accessible from the rest of the city. Some of the site has been reclaimed by rubbish tipping and the methane extractors delimit the early parts of the tipping site. The later parts of the site are still being tipped (28.3ha) and landscaped into 'hills'. The city and the surrounding districts have had problems both finding new acceptable landfill sites and locating a new waste incinerator to comply with modern emission quality requirements. The marina itself is the home of the international marine recreation company, Sunsail. The facilities are limited by the tidal nature of the upper harbour, confining access to high tide and the marina struggled initially.

The North Harbour site forms part of the northern boundary of the Portsmouth Harbour Special Protection Area (SPA) and Site of Special Scientific Interest (SSSI). It is also an EU Ramsar (wetland site) for the protection of the wildlife.

Figure 12: Former military land.

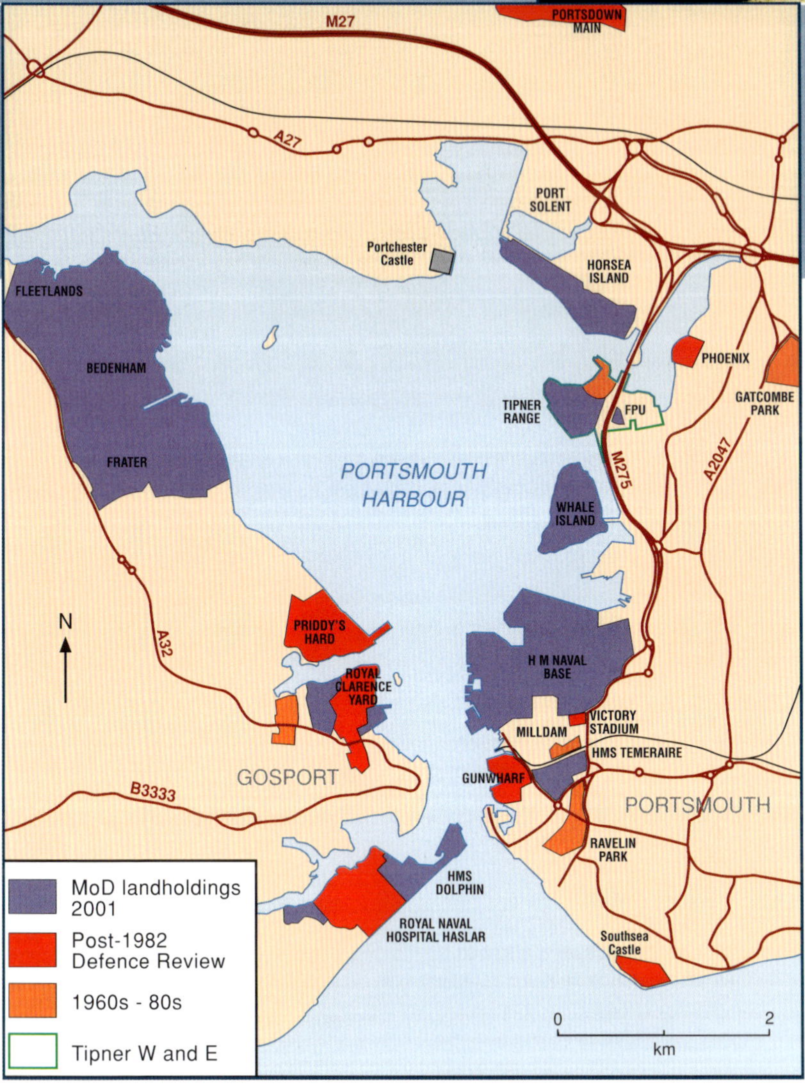

From militarism to consumerism

Military dominance of the city has been progressively reduced since 1945 and its spatial decline invites geographical investigation. Each phase and scheme is a response to ideological shifts, governmental policies and planning wisdom at the time of its redevelopment. Changes may be analysed from the perspective of the scale of forces impacting on the areas undergoing change. There are international, national and local forces at work shaping the developing townscape and its social geography. The process of redeveloping military lands began before the last decade.

Retreat from spatial dominance

We have seen how changed political relations with France and the advance of military technology resulted in

Former Clarence Barracks, now City Museum and Archives.

Palmerston's follies. In post-1945 UK it was not long before the role of the military was reassessed in the light of atomic weapons, new missile technologies and the declining global role of the UK. Reassessment of defence needs by successive governments has resulted in a trickle of land into new uses (Figure 12).

Brownfield developments:
The rise of public-to-public transfers

The initial releases of land in the 1960s involved the disposal of barracks and areas of open space owned by the MoD. Manpower was less important in modern warfare and married quarters were replacing the traditional barracks.

Clarence Barracks, built on the open land and defensive works of Old Portsmouth, was converted to become the City Museum and Archives. The adjacent Victoria Barracks was also released in the 1960s and demolished to be replaced by Pembroke Park, a private housing and apartment development socially in keeping with adjacent Old Portsmouth. A modern hotel, the first since 1909, was also constructed. Both of these uses were symptomatic of the local steer to diversify the city's economic base, with new industrial and office developments such as Zurich Insurance and IBM located in Portsea and the northern suburbs respectively.

Southsea Castle, a wartime army camp, was returned to civilian use in 1960. It became the site for a city-administered Museum. The grounds of the camp were redeveloped for the D-Day Museum.

Cambridge Barracks (1856), at the northern end of High Street, has been used by Portsmouth Grammar School since 1926. Nearby, Ravelin Park was part of the lands freed up when the medieval defences of Portsmouth were demolished. It remained in military use until 1973 when the University acquired the land, including Montgomery's wartime residence, Ravelin House. The NAAFI (the 'other ranks' social club) at the southwest corner of the park became the Students' Union. Much of the park remains as an open space surrounding the Library and a new Union. Outside the park's northeast corner, The Nuffield Club, constructed for officers in the 1950s, was also taken over by the University as a social facility. It was a similar tale for the conserved eighteenth and nineteenth-century Milldam Barracks which the University's Faculty of Humanities now occupies. (Milldam House was converted to become the local Registry Office.) In all of these cases one public use has replaced another, and the nature of the replacement was very much in keeping with the rise of Higher Education as a national priority with the designation of polytechnics as 'new'

Gunwharf Quays.

© Portsmouth City Council

universities. In Gosport, St Vincent Barracks became a Sixth Form College.

Fort Cumberland was constructed in the eighteenth century to defend southeast Portsea Island and the entrance to Langstone Harbour. It contains the best example of defensive architecture of the period. It was relinquished by the Royal Marines in 1973 and managed by English Heritage. Since 1997 when it was re-opened, it has become English Heritage's archaeological research establishment.

Elsewhere, Hilsea Barracks, once occupied by the army, was released for housing between 1971 and 1973, and the adjacent Garrison Recreation Ground was sold in 1982 to be retained as public open space and playing fields. Tipner Barracks was released and became a mixed industrial area.

Public to consumerist

The last twenty years of the century saw a shift in the ideology of redevelopment. The scale of land releases increased following the 1981-82 Defence Review and the post-1989 'peace dividend'. Sites that might have been redeveloped were refurbished by the private sector, while others were totally redeveloped in search of profit.

An example of refurbishment is the 12ha site of the 1865 Royal Marines Barracks at Eastney. Much more prominent have been the releases of land for major private investments. Horsea Island, which had housed a torpedo testing range, will form the southern boundary of Port Solent when it is released. Subsequently part of HMS Phoenix, the naval nuclear, biological and chemical training centre and fire fighting school whose main base was across Tipner Lake, was cleared of pollutants and sold for housing.

Victory Stadium, a naval athletics facility, was closed when the HMS Temeraire sports ground was developed. It was sold as a retail park and associated car park. Victory Retail Park has five users and suffers from access problems and proximity to the city centre. However, the space-extensive users represent a consumerist redevelopment of the site, in marked contrast to previous redevelopments.

In 1985 the Portsmouth Naval Heritage Trust was created to manage the post-1982 rationalisation of building uses following the Defence Review. One key local input to the post-1989 'peace dividend' planning was the Portsmouth Harbour Plan (1989, revised 2000) and the associated Renaissance of Portsmouth Harbour Millennium project. These formed part of a proposal to open up the harbour frontage to the public and, at the same time, provide business and tourism opportunities, thus embracing national policies for the

© David Burtenshaw

© University of Portsmouth

Figure 13: Gunwharf Quays.

Millennium to serve local regeneration needs. The proposal was co-ordinated by the Fareham, Gosport and Portsmouth local authorities. The project involved the latter two authorities, the Royal Navy, various Heritage Trusts and a consortium of public and private sector organisations within the South East Hampshire Business Partnership (SEHBP).

By far the most expensive and controversial scheme on former defence land has been the redevelopment of the 16ha HMS Vernon site, at the entrance to the harbour, into Gunwharf Quays (Figure 13). HMS Vernon was a naval torpedo school and a base for motor torpedo boats. It had been bombed during the war and its post-war facilities contained a mix of soulless office blocks and a set of buildings dating from the early nineteenth century. This was not the only waterside site released at the time. Royal Clarence Yard (16.77ha) and Priddy's Hard (40.5ha) on the west side of the harbour were also declared surplus to requirements.

The planning brief required the conservation of the Old Customs House (Vernon Building) (1811), the Vulcan Building (an armaments store, 1811-74), and the Ariadne Building together with the Gatehouse and wall surrounding the site. Part of the brief also included the Millennium Tower, a proposed 150m tower at the entrance of the harbour to celebrate the Millennium. The tower was to be financed primarily from National Lottery Funds. The contract for the development, including the tower, was given to the Berkeley Homes group and the northern half was to be developed by Lordland Holdings, a company who had successfully regenerated the Victoria and Alfred waterfront in Cape Town. Both international and national influences were coming to bear on this development despite its avowedly local aims.

The advantages of the £150 million development are that it:

- Is close to the Heritage Area of the dockyard;
- Is close to The Hard transport interchange with its rail and bus connections and ferries;
- Has a mix of facilities that will attract the young and old alike;
- Employs 2000 people, many part time;
- Is visible to tourists from the ferries from France and Spain; and
- Has ample car parking.

Shopping at Gunwharf.

© Portsmouth City Council

The disadvantages of the scheme are:

- Road access through the city is not easy, especially at peak times and peak seasons;
- The housing is in a clearly segregated area;
- It is cut off from the relatively deprived area of Portsea to the north;
- It provides no local facilities either for the residents or for Portsea residents. In the latter case they still have a 1km journey to the essential shops;
- It provides few jobs for the people of Portsea and more for the ephemeral student population of the city;
- The proposed hotel has not materialised; and
- The Ariadne Building, which was scheduled for university use, was demolished because the University was unable to meet the costs of occupying the building.

The saga of the Millennium Tower (see Trail 1, page 39) illustrates not only the issues that surround the whole process of obtaining rights to develop sites in Portsmouth and elsewhere, but also the power that major commercial groups have over local councils. They have the ability and the cash to be able to use involvement in developments such as the tower to manipulate decision makers. The developers have gained the right to make a profit from the scheme partly by breaking a promise to the people of the city. The council is then blamed in the press for the consequent 'shame' brought on the city by its councillors who approved the bid. This is a classic case of 'who gains and who loses?' – the focus of welfare geography.

The gainers are:

1. The developers who gain from rents and leases;
2. The investors in the developers who might be the pension groups subscribed to by people in Portsmouth: 'Gunwharf offers us the opportunity of income as well as providing good returns for shareholders', Ian Henderson, Chief Executive of Land Securities;
3. The owners of the housing;
4. The major international designer label companies;
5. Bar, club and food chains;
6. The cinema owners who have established successful outlets at Gunwharf;
7. The 500 unemployed people who now have jobs;
8. The local economy which benefits by about £30 million per annum; and
9. The taxpayers who saw money paid to the government for the MoD site in the first instance.

© David Burtenshaw

Converted hotels, Southsea.

© David Burtenshaw

The losers are:

1. The people of Portsmouth whose city has spent money negotiating for a tower – so far to no avail;
2. The people of Portsea who now look at the slab wall of the scheme;
3. Those who have to endure the extra congestion caused by the development being at one corner of the island, remote from its catchment population of 2.7 million people within 1 hour's journey time;
4. The 1500 jobs forecast in the original proposal which have yet to be created; and
5. Other retailers of similar goods elsewhere in the city.

A similar issue to Gunwharf can be seen in its early stages at Tipner. Tipner is an area of mixed usage whose present state is the product of military use and abandonment, and national transport policies. The construction of the M275 divided the area and has increased its potential while creating problems for any future development. It also drew attention to the run-down appearance and visual intrusion of the area as the motorway crosses into Portsmouth. There has been a succession of proposals for the area, all of which have failed to reach fruition.

Other sites in the city region

HMS Daedalus (a naval air station) at Lee-on-the-Solent was also scheduled for closure. Inland, HMS Mercury, a communications training school near Petersfield, was also declared surplus. There are three further former military sites in Gosport that are being marketed for housing, commercial and employment uses. The next challenge is the return of Portsdown Main (the former Defence Research Agency establishment on the crest of Portsdown Hill).

Tourism: from resort to defence heritage

In the nineteenth century Southsea grew both as the officers' settlement and as a seaside resort. In common with most UK resorts the stimulus was the building of the railway. Since 1945 there has been a gradual transformation from a traditional tourist resort to the modern, heritage dependent city of 2002.

Traditional resort

Southsea is where the greatest number of resort hotels and boarding houses can be found. Many of the larger houses, either on the seafront or in the roads immediately behind, were converted to hotels and bed & breakfast establishments. From these bases the tourist was able to enjoy the traditional seaside activities whose attractiveness was enhanced by the wide expanse of Southsea Common. Typically, these activities included walks on the promenade and in the gardens, swimming from the pebble beach, a visit to the fairground, boat trips and evenings at the

Figure 14: Heritage tourism.

seaside theatre. Contact with the military function was limited to watching vessels arrive and depart, visiting HMS Victory, and the annual Navy Days. Periodically there was the more spectacular Review of the Fleet, coinciding with coronations and jubilees. People were encouraged to visit Portchester Castle or cross to the Isle of Wight for day trips. All of this came under threat in the 1960s: the advance of the package holiday abroad saw Southsea move slowly into decline; day trips replaced the week's visit and the face of the resort became dated.

Military tourism

The triggers for an upsurge in tourism were threefold:

1) The opening of the Continental Ferry Port in 1976 with the potential to attract foreign visitors,
2) The Defence Review, which stimulated consideration of alternative uses for redundant space, and
3) The raising of the Mary Rose in 1982 and the need to find a place to exhibit the vessel and its artefacts.

Underpinning all of these was a policy shift away from redevelopment to refurbishment and conservation of heritage. Throughout the 1970s the classic housing areas of Southsea, such as Owen's Southsea, were subject to conservation orders (Charles Dickens'

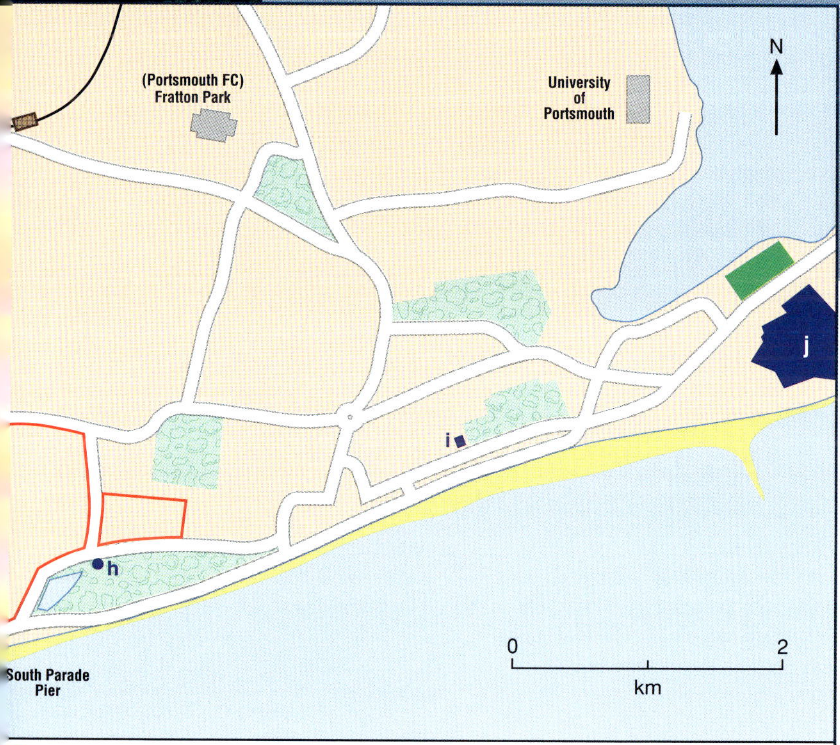

d Sea Life Centre	g Skatepark	j Fort Cumberland
e D-Day Museum	h Aquarium	
f Southsea Castle and Museum	i Royal Marines Museum	

birthplace is in a conservation area formed in 1970). Only Eastney Barracks (1973) was subject to conservation orders and the rest of the military heritage had to wait until 1981 when 52.4ha of the naval base and adjacent Portsea were conserved.

Several business/tourism hotels were constructed on routes into the city and at the ferry port. As with hotel construction, disparate developments gradually merged into a coherent strategy which reached its culmination at the turn of the century. All of the following capitalised on British naval and military history associated with the city;

- The housing of the Mary Rose in the Dockyard;
- The improved Royal Navy Museum in the Dockyard;
- The restoration of the first submarine, Holland 1, and the installation of HMS Alliance in the Royal Navy Submarine Museum housed in the former submarine base, HMS Dolphin, Gosport;
- The arrival of the restored HMS Warrior (1860), the first iron-clad battleship;
- The development of the D-Day Museum and Overlord Embroidery; and
- The opening of the Royal Marines Museum at Eastney.

Figure 15: Spitbank Fort.

© David Burtenshaw

Heritage tourism

A common theme that emerged in keeping with the ideology of the 1990s was heritage tourism. Under this banner it was possible to envisage tourism attracting more overseas visitors to bolster the flow of recreational (day) visitors.

The catalyst for unifying the new and the proposed developments and for reinvigorating the older tourist environment was the Millennium together with the development of National Lottery Funding for major heritage projects. The successful £40 million Millennium Landmark bid for the Renaissance of Portsmouth Harbour project has created the largest tourism attraction on the south coast of England.

The physical links between all of the projects are the two Millennium Promenades. Old-fashioned Navy Days have been replaced by Tri-Service activities linked to the three-day International Festival of the Sea. This event ties together local history and pageantry, national defence exhibitions and public relations (e.g. Defence 2001 and the presence of tall ships from around the world). The sponsors are all major transnational companies such as Rolls Royce, Alenia-BAE Systems and BP. Such is the popularity of the event that 15,000 park-and-ride places are set up beyond the city. City tourism's global status is reinforced.

Tourist literature tries to reinforce Portsmouth Harbour's new global status by referring to a 'World Class Waterfront'. The range of heritage sites, Gunwharf, the traditional range of resort activities along Southsea seafront, an Aquarium, a covered Leisure Pool facility, the D-Day and Royal Marines Museums, all provide a 6.5km linear resource (Figure 14). These separate tourist honeypots perform different functions which can be examined in the field. It is possible to distinguish international, national or local provision; day- or night-time activities; age and indeed gender targeting; and impact.

Further afield Port Solent is also marketed as a part of the World Class Waterfront. The Royal Armouries have a museum in Fort Nelson on Portsdown Hill and Spitbank Fort, one of Palmerston's offshore follies, can be used for conferences, corporate entertaining and social events (Figure 15).

© Portsmouth City Council

Royal Armouries, Fort Nelson.

© Board of the Trustees of the Armouries

The benefits of tourism

1. In 1999-2000 half a million visitors to the Dockyard Heritage Area brought approximately £2.5 million to the local economy.
2. Waterfront developments are estimated to create 3500 jobs and to boost visitor numbers from 1.6 to 6 million per year. Publicity from the proponents of the Millennium bid suggest that a potential extra £50 million will be pumped into the local economy.
3. As facilities become known, more people will come to the city not only from the UK but overseas. In 1996 there were 4.2 million visitors and 4.7 million in 1998. The projected figure for 2002 is 6 million.
4. It might serve to attract other business to the area and students to the University.
5. There is an employment multiplier.
6. The image of the city is enhanced, a process which has been termed 'urban boosterism'.

The challenges for tourism

1. The Channel Tunnel could siphon cross-channel visitors away from the city in the long term.
2. The emphasis on military heritage might not appeal to future generations and diversification might be necessary.
3. Access to the major attractions is not easy by road at present and parking is limited. The proposed permanent development of park-and-ride (e.g. Tipner and Port Solent) and Light Tramways might help.
4. Are the other facilities, such as hotels and accommodation, of the appropriate standard? Big hotels are planned for but no group is showing much interest.
5. Does the city have an image problem resulting from occasional adverse publicity in the national media?
6. The possibility of leakage of income back to major multinational companies, e.g. Starbucks, Gap.
7. The decline in the number of small hotels, evident only when major events create demands exceeding supply. Between 1996 and 2000, 46 hotels gained permission to convert to luxury homes, hostels and holiday homes.

Small area studies and trails

Contents

© University of Portsmouth

Trail 1: Heritage trail 37

Trail 2: Southsea trail 42

Trail 3: Tipner 45

Trail 4: Gosport heritage trail 48

© David Burtenshaw

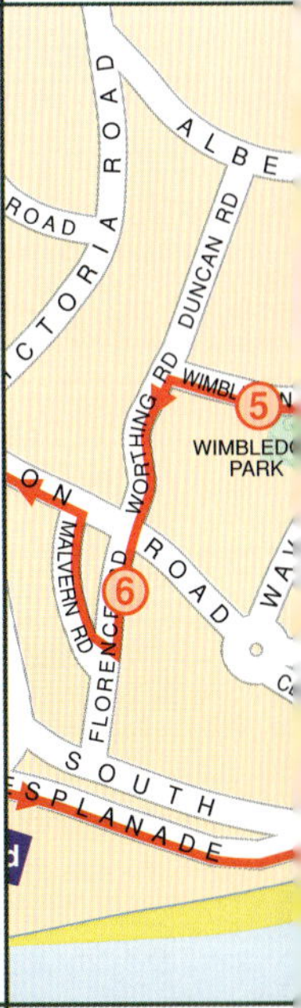

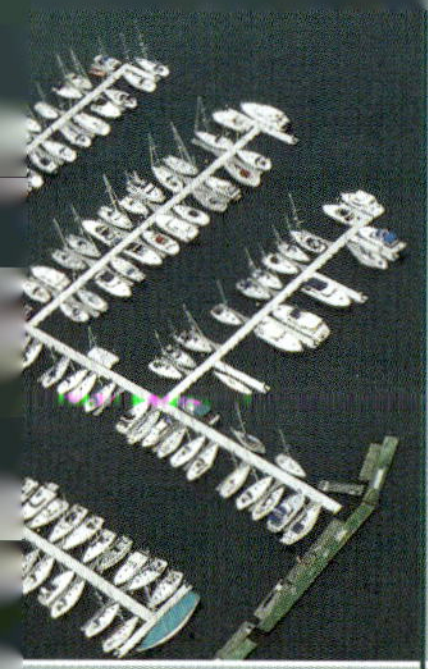

© Camper & Nicholsons Marinas Ltd.

Cascades Centre

© Taylor Woodrow Property Company Ltd.

Introduction

The trails are designed to illustrate the themes of the book: The heritage trail enables observation of the oldest districts of the city, their naval heritage and use of former marshland and defence land; the Southsea trail examines the contrasting social areas and the impact of the physical geography on the resort; the Tipner trail explores a classic, polluted, brownfield site at the entrance to the city; finally, the two Gosport trails examine the reuse of military land on the less popular side of the harbour.

Other themes of potential interest are:

Retailing issues

Gosport, Gunwharf and Southsea shopping centres feature in the Trails. A fourth centre, Commercial Road, is the city centre. The centre is pedestrianised and includes the Tricorn (1964) (voted the UK's ugliest building by Radio 4 in 2001 and now awaiting demolition) and a covered centre, Cascades. Commercial Road exhibits classic clustering of activities, especially of quasi-retailing to the south. It is possible to contrast all four centres using Goad Plans. Why is each distinctive? Can all four survive?

Sailing issues

It would be possible to analyse the different marinas and sailing clubs around the harbour in terms of their facilities, access by land and sea, and whether membership is local or national.

Leisure issues

Competition between South Parade Pier, Central Southsea, Guildhall Walk, Port Solent and Gunwharf Quays for night-time entertainment.

The changing role of Portsdown Hill

The contrasts of north and south, east and west are worthy of investigation together with the variety of uses and relic uses.

Trail 1: Heritage trail

Distance: 10.5km
Walking time (without stops): 2 hours
Disabled access: Yes

1. Starting from the university car park in St James Street, turn left into Queen Street and continue through the heart of Portsea. Curzon Howe Road is a conservation area of interwar housing and part of the Dockyard Conservation Area. At the end turn right through Victory Gate into the Historic Dockyard.
2. The route ahead leads to the Mary Rose Ship Hall (former No 3. Dock),

Trail 1.

a University of Portsmouth	g HMS Warrior	m I.O.W. Ferry Terminal
b Mary Rose Ship Hall	h Transport Interchange	n Spice Island
c HMS Victory	i Portsmouth Harbour Station	o The Round Tower
d Royal Naval Museum	j Vernon Building	p Cathedral
e Mary Rose Museum	k Millennium Tower	q Grammar School
f Boat House 6	l Vulcan Building	r Old Milldam Barracks

HMS Victory (former No. 2 Dock) and the Royal Naval Museum passing through much of the architectural heritage including the 334m long Great Ropery (1776) on the right hand side. Retrace your

© University of Portsmouth

© University of Portsmouth

route and on the left are the Mary Rose Museum and, across the 'pond', Boathouse 6.

3. Boathouse 6 (1848) has been restored to a dual role: 'Action Stations' is a state of the art interactive experience focusing on the modern navy. The Institute of Maritime and Heritage Studies has been established by the University of Portsmouth with local and national sponsorship to develop courses that can build on the wealth of architecture, historic ships, archives and library resources in the Historic Dockyard. The Institute is a founder member of ReNDoc, a European network of dockyard towns, maritime museums and research institutes which should draw in researchers and visitors from the Baltic to the Mediterranean.

4. Back through the Victory Gate follow the Millennium Promenade which links the Historic Dockyard to Gunwharf and Old Portsmouth, identifiable by its distinctive street lighting and pavement motif. HMS Warrior is on the right hand side. The Hard is now a transport interchange for rail and buses, and for passenger ferries to Gosport and the Isle of Wight. Retailing here was once related to the dockyard, although the hotel is a last vestige of better times. Continue along the Hard until you come to a passage under the railway leading to Gunwharf Quays.

5. Gunwharf was identified in the Portsmouth Local Plan (1995) as suitable for mixed-use development. The approved scheme involves two distinct halves. The northern half has been developed as a 50,000 sq. m commercial venture which

Bars at Gunwharf.

© David Burtenshaw

opened in 2001. It comprises 65 shops, 25 bars and cafés fringing the southern and western waterfronts, a BowlPlex and an 11-screen cinema. There is space for occasional markets and displays. Beneath the scheme is parking for 1400 cars. It is already possible to identify the less profitable corners from pedestrian flows, the quality (and/or brand name) of the products sold and the distribution of empty units. There is also a contrast between the daytime retail environment and the night-time clubs and pubs atmosphere that is proving successful, especially in the summer months. The sole conserved building, The Old Customs House, has been converted into a public house. Gunwharf is also a maritime venue. It has a number of boat moorings and is a focus of events such as the International Festival of the Sea. It also provides facilities not available in the city and only found in an out-of-town development between Fareham and Southampton. However, one has to ask whether the city can maintain three centres if the flow of visitors reduces significantly.

The southern, residential part is under construction (2002) with a mixture of town housing furthest from, and luxury apartments closest to, the waterfront. The Vulcan Building is being converted to apartments with a new replica wing on its northern side. The scheme is an undoubted success if the speed with which the 310 properties have been bought ahead of completion is a guide. The area is clearly segregated from Old Portsmouth by a wall and from the commercial area of Gunwharf by the relic of the Mill Pond conserved as a linear water feature through the site.

Perhaps the most controversial aspect of Gunwharf has been the Millennium tower. The Berkeley Group included the development of the £28 million tower in their planning proposals to the city. The public vote was in favour of a spinnaker-shaped tower with a viewing platform – a visible landmark from up to 40km away. No sooner had Berkeley Homes obtained the rights to develop the site than they withdrew from the tower project. They maintained that it was uneconomic. Also, the Berkeley Group sold half of their interest in Gunwharf for £80 million in December 2000, so realising a substantial part of their investment in the whole scheme. Many now say the Berkeley Group never had any interest in the tower. Their interest was in obtaining the rights to develop the site and obtaining

© David Burtenshaw

© University of Portsmouth

Spice Island.

the maximum benefit for their shareholders.

Subsequently, efforts were made to include other features in the tower design, including a freefall ride. However, companies were anxious not to take risks investing their money alongside that of the National Lottery and the city. The project is still progressing, although at the time of writing the tower site remains a semi-circular patch of foundations at the northwestern edge of the scheme, fringed by the majority of the empty retail sites.

6. Leave Gunwharf, turning right into Gunwharf Road and passing the Isle of Wight car ferry terminal. Gunwharf Gate was redeveloped on the former power station site for a residential extension to Old Portsmouth in the 1980s. Continue along White Hart Road passing the Camber fish dock.
7. Turn right into Feltham Row footpath. On the left is housing that replaced Vosper's ship building yard. Turn right into Broad Street which passes onto 'Spice Island', a low-lying spit at the mouth of the harbour. Today, this area is a mixture of residential and recreational uses, the latter under threat from Gunwharf. Towards the end of Broad Street a car park and unused buildings on the right await development.
8. At The Point turn left and retrace your route through Bath Square and West Street. Climb up onto the old defences of the Round Tower for good views across the area, follow the Hot Walls and descend before the Square Tower.
9. Walk north up High Street, the former commercial heart of the city, past the cathedral. The street is a mixture of conserved dwellings and modern infilling to replace bomb damage. The Grammar School on the right is housed in the former Clarence Barracks, behind which, on Museum Road, is the City Museum and the former NAAFI which was a Students' Union until 2002.
10. Continue on to Cambridge Road. Ravelin Park to the right, once part of the town's defences, is now part of the University.
11. Turn left into Burnaby Road, which traverses the naval recreational facilities sited on the old Mill Pond. Under the railway bridge, pass the former Milldam Barracks on the left, also now part of the University, and round the Portland Building to return to the start of the trail.

Trail 2.

Trail 2: Southsea trail

Distance: 4.4km (8.2km with both detours)
Walking time (without stops): 1 hour (1 hour 45 minutes with detours)
Disabled access: Yes

1. Start from the car park alongside the D-Day Museum (1984) and follow Clarence Esplanade east past The Pyramids Centre and Southsea Castle along the beach ridge towards South Parade Pier. Looking across Southsea Common and the gardens, the few remaining traditional seafront hotels on Clarence Parade are clearly visible. The large white building facing southwest towards the skate park was once the Pendragon Hotel and is now luxury flats. Near the Pier, besides the traditional pier activities there are nightclubs and two larger chain hotels, one of which has converted space to apartments. This area is also threatened by the developments at Gunwharf.

2a. Cross the Esplanade and turn right along St Helen's Parade with Canoe Lake on your right. Turn left into Craneswater Park.

2b. The following part of the route may be left out as it adds a 3.4km loop to the trail. Continue east along Southsea Esplanade past Canoe Lake and Lumps Fort. Eastney Barracks, the nineteenth century, Grade II listed, barracks (used in the film *Cockleshell Heroes*) was redeveloped as a complex of 67 high-quality apartments that have been occupied primarily by middle-aged, newly affluent people. There were plans to convert part into an hotel but these were dropped because the location was inappropriate even though the building was ideal. The new housing developed on the north of the barracks is without views to the Solent and is certainly not of the same standard of interior opulence as the apartments in the converted

© City Museum and Records Office

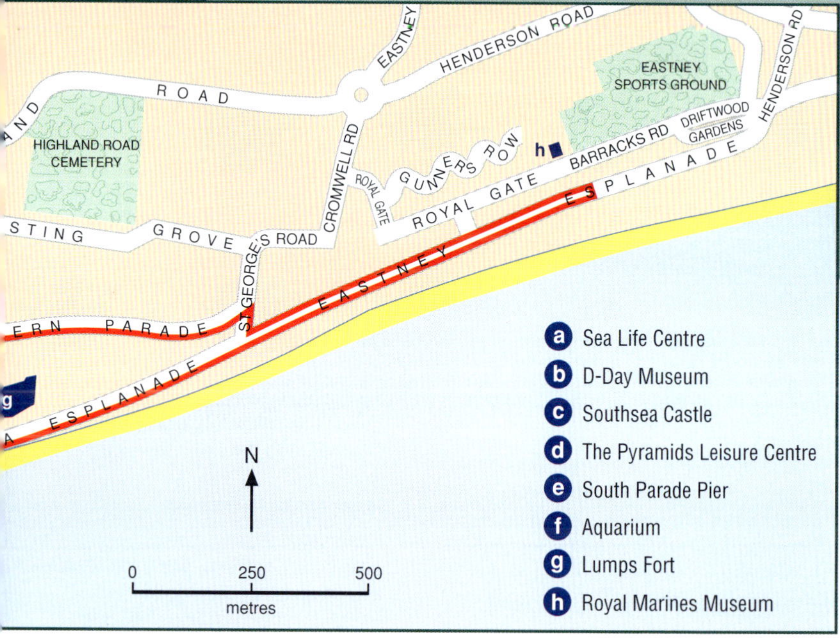

barracks. At the eastern end is The Royal Marines Museum. Return via St George's Road and Eastern Parade to rejoin the trail at Craneswater Park.

3. Craneswater and the roads off Eastern Parade are a typical mix of apartments, large houses and small hotels. East Southsea to the east of Festing Road is not part of the new town council area. Take the next left and left again into Craneswater Avenue, turn right into Old Bridge Road and right again along St Ronan's Road. The bridge traversed the old Southsea railway line and the road follows its course towards Fratton.

4. Turn left into Albert Road, a linear retail street providing facilities for the densely populated districts on either side. The road is the heart of student Southsea – note the lettings agency as you turn into Albert Road. Many students live in the terraced streets to the north.

5. At the traffic lights turn left into Waverley Road and take the third right into Wimbledon Park Road. At this point you are at the most northerly point of one arm of the Great Morass, which was drained and building had begun on the western fringes by 1870 (see pages 23-25). Modern redevelopments include housing and St Swithun's School.

6. Turn left into Worthing Road then cross into Florence Road before turning right into Malvern Road and left into Clarendon Road. This takes you through the main Morass area. The open spaces and most modern developments still mark out the former stream courses.

The Morass almost returned to its former state on 15 September 2000 when severe storms overwhelmed the city's sewage pumping station. Inspection covers were forced into the air by the water pressure and storm water and sewage flooded into the streets up to 1m deep. The floods damaged 750 homes in the area, a small school and several businesses such as the Florence Inn.

Southsea.

© University of Portsmouth

Clarendon Gardens, the lowest point in the Morass, became a lake up to 1.5m deep. The insurance and Council clean-up costs amounted to £2.25 million.

'The fountains pushing up through the drains were such an amazing sight and the waters rushing down the street instead of traffic were at first simply astonishing. Then the seriousness of it hit us. People at the lowest point of Florence Road lost everything covered with stinking sewage mess. Over the next weeks skips of sodden carpets, furniture and treasured possessions were filled. How dare Southern Water say it was an act of God!' (Adapted from *Portsmouth Society News*, Autumn 2000)

This event can be debated. Was it 'An act of God' or did the fault lie with our forefathers who drained and built on the Great Morass? Does Southern Water have any responsibility for the floods or will the costs now be transferred to higher insurance premiums in the area? The area, which had become notorious for subsidence as the Morass dried out during warm summers in the late twentieth century, is notable today for the obvious signs of redecoration following flooding. The tell-tale signs of subsidence are less easy to spot, except in Somerset and Beach Roads.

7. Follow Clarendon Road and turn left into The Vale. Lymington Lodge, on the corner of The Circle (1852), and The Vale (1850) are distinctive parts of Thomas Ellis Owen's Southsea, which was developed to house the officer classes. Return to Clarendon Road and turn right by Knight and Lee into the Palmerston Road shopping precinct.

8. Palmerston Road is the shopping centre for Southsea, although it has a wider catchment due to the presence of two of Portsmouth's department stores. Following wartime bombing there was much rebuilding to restore a more genteel and middle class shopping area for the inhabitants of Southsea. However, it is having to compete both with the larger Commercial Road shopping area and also with Gunwharf Quays.

9. At the northern end of the precinct turn left into Kent Road. From here it is possible to detour 0.9km to Queens Crescent (1847) and Sussex Terrace (1854-55).

10. Turning left into Portland Street, pass Portland Terrace (1846) before turning left into Osborne Road and right into the lower half of Palmerston Road. South of the precinct the facilities have declined, with the exception of quasi-retailing (banks and estate agents). There are many cafés, restaurants and clubs in this area because it has been planned as a recreational zone, much to the dismay of Southsea Council.

Continue south along Avenue de Caen back to the start of the trail.

Trail 3.

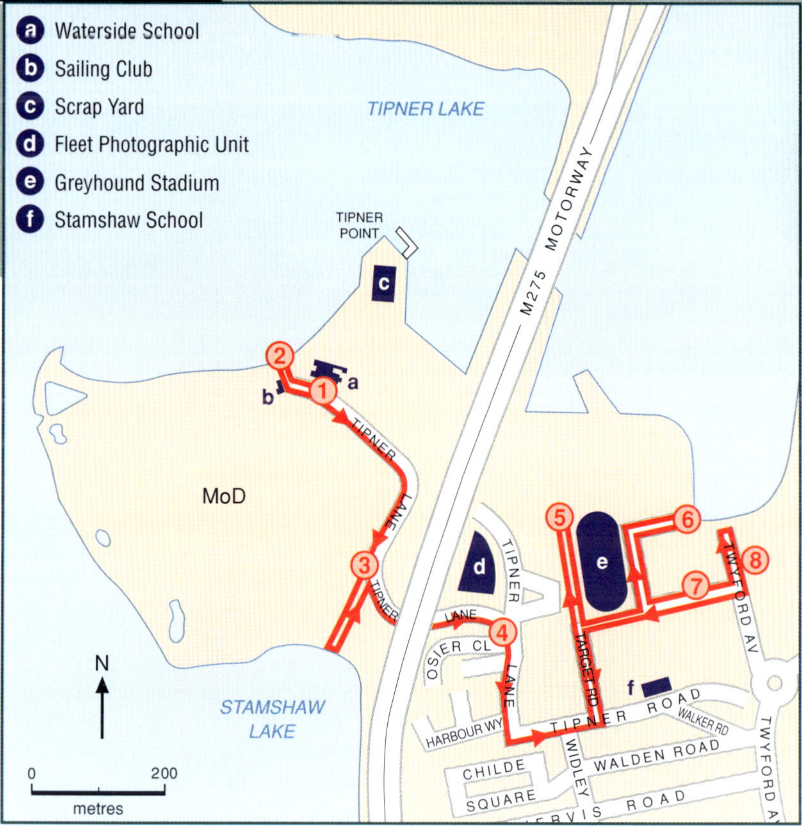

Trail 3: Tipner

Distance: 3.6km
Walking time (without stops): 45-60 minutes
Disabled access: Possible but rough ground at several points

Introduction

The advantage of the Tipner site is that it straddles the motorway leading to Portsmouth. Some preparation has already been made for a junction which has not been completed. The site overlooks the harbour and is at the gateway to the city where any development will be prominent. It would also bring employment into a ward which currently has very little (77% commute out of the ward) and where deprivation is high. There is open space which could be retained and enlarged by creating a coastal path which would benefit the local community.

The disadvantages of the site are the contamination, its multiple ownership, the constraint of the firing ranges, the noise of the motorway, the proximity of the SSSI and Special Protection Area (SPA) in Tipner Lake – the habitat of 120 plant species including Bee Orchids – the lack of a motorway junction and the problems of access through an area of closely packed terrace housing. There is also the question of what to do with the Greyhound Stadium.

The Tipner community to the south is the third most deprived in the city and has the second lowest income per worker from a workforce which is

Tipner.

© University of Portsmouth

24.5% skilled manual compared with a city average of 3.3%.

The following are among the proposals for parts of the site that have been made over the past three decades:

1. A 9000-capacity arena and associated businesses.
2. An office park with retailing.
3. A large-scale, out-of-town retail development (such as IKEA).
4. Housing on the seaward side.
5. Some small-scale reclamation of Tipner Lake.
6. Retain the school and firing ranges.
7. Relocate the ranges and FPU to other defence owned sites.
8. Hotel and Conference centre.
9. A Park-and-Ride terminal for up to 900 vehicles and a bus link to the city centre and Gunwharf.
10. A Park-and-Ride terminal but with a tram stop on a link from Port Solent to the city centre.
11. A cycleway using the bridge for no. 10 above, together with a coastal cycle track and footpath.
12. Retain the Greyhound stadium or relocate it elsewhere in Tipner.
13. Improved marine slipways and sailing facilities.
14. A lorry park and warehousing.

In addition, if any existing uses need relocating, alternative sites will need to be found.

In 2001, when three groups of 13-15 year olds from local schools studied the area in collaboration with the University, they all favoured some kind of open space and outdoor activity centres that would appeal to the teenage market rather than the more cash rich ideas of the developers and the city. The purpose of any visit to the area should be to establish what those who are underprovided for might want, be they the youth of the city or the inhabitants of Stamshaw and Tipner.

1. The trail starts near Waterside School. Walk out along the open land onto the foreshore. Much of this western half is used by Tipner Range (MoD).
2. Beyond Lion, the sail training establishment, the outer reaches of Tipner Lake can be seen. Portchester Castle is to the northwest and Horsea Island and the waste tips are across the lake. From the foreshore land use of the former Barracks on Tipner Point can be seen (ready-mixed concrete and scrap yard dismantling naval vessels). The scrap yard once extended to the east of the motorway, but the low bridge gave

© David Burtenshaw

Portchester Castle.

© English Heritage Photo Library

the site access problems. Closer to the school is a sailing club relocated from the present course of the M27.

3. Walk back along Tipner Lane. The reclaimed field on the left after the school is known to have been built up by dumping topsoil over ponds where, until 1956, the military deposited munitions waste. Turn right beside the Angling Club and go to the edge of Stamshaw Lake. This yacht and angling club does not enjoy good water access because of the tidal range and the effects of the M275 motorway reducing tidal scour.
4. Return along the cycle track and pass under the motorway; this is the southern half of an incomplete junction. The Fleet Photographic Unit is beyond the small field on the left.
5. Turn right into Tipner Lane. Follow the road round and turn left into Tipner Road. The route takes you through typical housing for the area, stretching south towards the city centre. Turn left into Target Road. To the left of the Greyhound Stadium (1870s) it is possible to continue up to a small bank overlooking the former Harry Pounds scrap yard (made famous at the end of the film *Tommy*), now overgrown with bushes.
6. Return to the southern end of the stadium car park, turn left and follow the perimeter along a dirt road leading to the entrance of the former scrap yard. Do not enter the site without permission because it is highly polluted. Soil samples taken on the site reveal lead pollution well above acceptable levels.
7. Return back down the access road past a host of small building and storage uses on the site of a former coal depot and the ready-mixed concrete depot. Turning left, Stamshaw School can be seen to the right. The school had to have special membranes laid under its playing field to combat pollution from infilling in the past.
8. Turn left and walk to the end of Twyford Avenue, from where the inner Tipner Lake can be seen together with the Jack Tar sculpture providing a modern bird roost. At low tide the problems of waste dumping can be seen here. The Mountbatten Sports Centre is to the east.

Return to Tipner Lane via Target Road and back to where the trail started.

© Camper & Nicholsons Marinas Ltd

Trail 4: Gosport heritage trails

Distance: Trail 4a: 2.7km
Trail 4b: 4.9km
Walking time (without stops): Trail 4a: 30 minutes; Trail 4b: 50 minutes
Disabled access: Yes, with exception of HMS Alliance

This trail has two loops starting from the ferry terminal. It can also be started at Priddy's Hard or at the Royal Navy Submarine Museum. Much of Trail 4a is along the second Millennium Promenade, which links the Submarine Museum, the Gosport ferry terminal, Royal Clarence Yard and Priddy's Hard.

Trail 4a

1. On leaving the ferry turn left along the promenade. On the right is an area of social housing developed in the 1960s. From the pier at Haslar Marina there are good views of the whole of the Portsmouth Dockyard, Gunwharf, Old Portsmouth and, through the harbour mouth, Clarence Pier and Southsea Castle.
2. Haslar Marina houses 620 boats and is the home of chartering companies. Before the end of the marina the promenade passes through some of Gosport's ramparts and a small defensive moat can be seen on the right beyond these. On reaching the road turn left by the Marina offices and cross Haslar Bridge which is effectively the upstream limit to yacht navigation.
3. All along the promenade and on the bridge good views of the 1905-94 submarine base, HMS Dolphin, can be seen on the left. HMS Alliance looms large over the Submarine Museum. To the left of the submarine the narrow width of the spit can be seen with a road running along its crest. The road leads to Fort Blockhouse which is the national centre for Tri-Serve Nurse Training, operated in collaboration with the University of Portsmouth.
4. At the eastern end of Haslar Bridge is the JSASTC (Joint Services Adventurous Sail Training Centre) – another joint military resource. At this point the road is lined by two high walls. On the left are the fenced grounds of the Royal Hospital Haslar, an MoD hospital which accepts civilians. Plans to close this hospital have met with violent opposition from people in Gosport who would have to make the long journey to Cosham to obtain similar facilities. On the right is QinetiQ (the new name for the Defence Research Agency, DERA) a Marine Technology Park for both the private and public sector.

© RN Submarine Museum, Gosport

Trail 4a.

Camper & Nicholsons Marina, Gosport.

© Camper & Nicholsons Marinas Ltd

5. Retrace your route but follow Haslar Road, passing the moat on the right. The route passes through Trinity Green and more social housing. Cross South Street into Gosport town centre. Gosport has always suffered from competition from Portsmouth and Fareham which both have greater potential catchment areas. Turn left, then right into Cross Street and right into Mumby Road. On the left side is Camper and Nicholsons Marina (for 320 boats) where Berkeley Homes have built more luxury waterside apartments. Continue along Mumby Road through the Falklands Gardens back to the ferry terminal.

Trail 4b

1. On leaving the ferry terminal cross the road and walk along the High Street before turning right into Cross Street and right into Mumby Road and Camper and Nicholsons Marina, as above.
2. At the entrance gate, cross the marina yard in a northerly direction to emerge on Harbour Road (also part of Millennium Promenade) and follow the road back to Mumby Road.
3. As you turn right into Weevil Lane, the redundant St George's Barracks South, with its shaded balconies designed for warmer climes, can be seen to the left. The barracks is due to be refurbished to create 55 flats, plus an additional 112 new dwellings, offices and community facilities to be built in the grounds. The course of the former rail lines to Royal Clarence Yard are still evident in the undergrowth.
4. Weevil Lane is the boundary between Royal Clarence Yard, the former stores and victualling (provisions) yard in use since 1704, and the former St George's Barracks North, which was closed in 1991. There are several listed buildings on the site. Berkeley Homes Group has been selected as the developer of a mixed-use, £65 million development, including 350 dwellings, leisure facilities and a sailing-events centre, to be completed by 2004. One drawback of this site is the fact that the MoD has retained

© QinetiQ

Trail 4b.

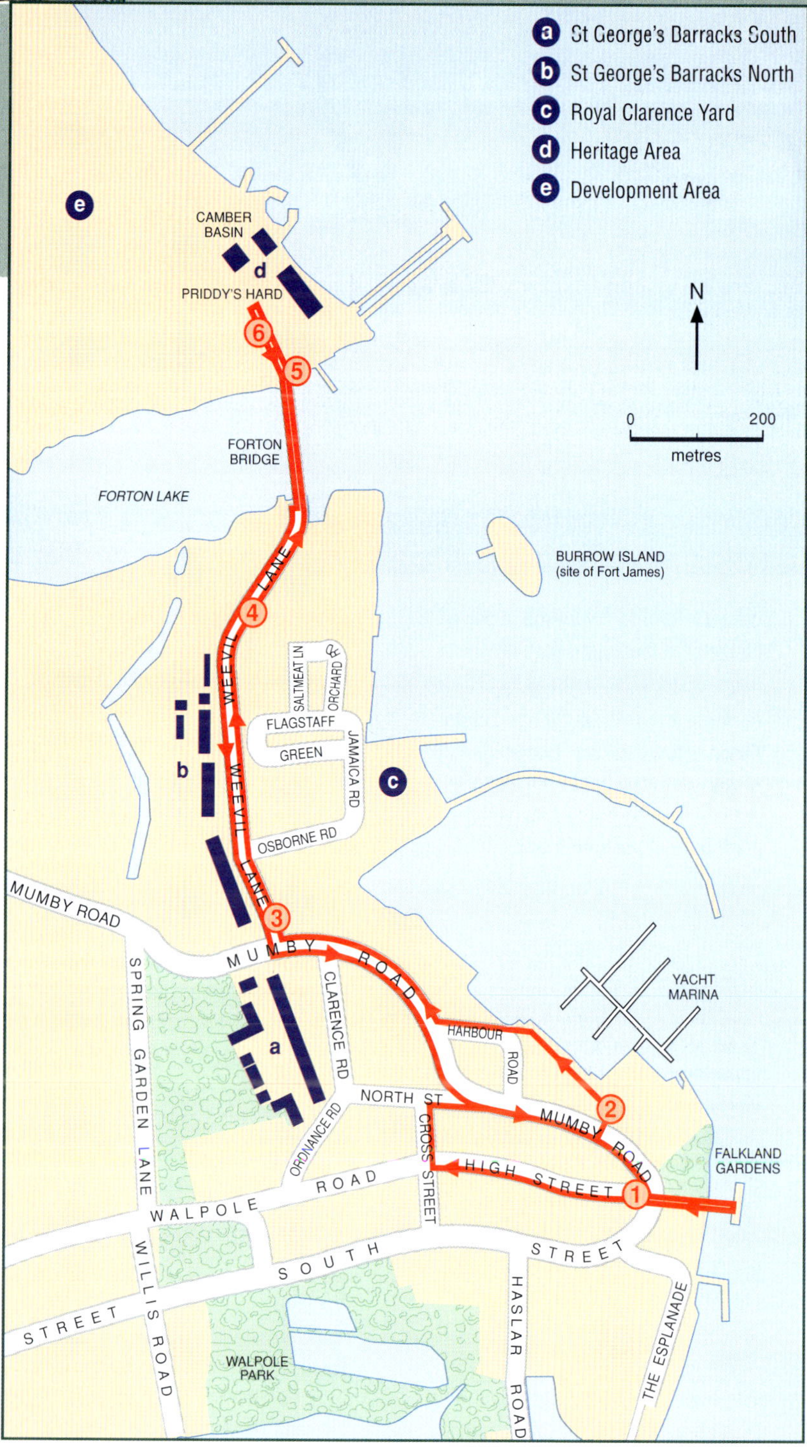
a St George's Barracks South
b St George's Barracks North
c Royal Clarence Yard
d Heritage Area
e Development Area
CAMBER BASIN
PRIDDY'S HARD
N
0
200
metres
FORTON BRIDGE
FORTON LAKE
BURROW ISLAND (site of Fort James)
SALTMEAT LN
ORCHARD
FLAGSTAFF
GREEN
JAMAICA RD
OSBORNE RD
WEEVIL LANE
MUMBY ROAD
SPRING GARDEN LANE
CLARENCE RD
YACHT MARINA
HARBOUR ROAD
NORTH ST
ORDNANCE RD
CROSS STREET
HIGH STREET
FALKLAND GARDENS
WALPOLE ROAD
SOUTH STREET
WILLIS ROAD
WALPOLE PARK
HASLAR ROAD
THE ESPLANADE

St Georges Barracks.

© Gosport Borough Council

© Gosport Borough Council

the tank farm to the west and the oil fuel jetty to the south, both of which might deter future occupiers.

5. Continue along Weevil Lane and cross Forton Lake by the new bridge. The bridge is the terminal for a shuttle bus from the ferry to Priddy's Hard on the north bank of the inlet. A small, more local, sailing club is located on the southeast side of the inlet.

6. Priddy's Hard was originally bought in 1759 and developed as a part of the defences of Portsmouth Harbour then, in 1768, as an ordnance depot. The isolated site surrounded by open space and water was ideal for its function as an armaments depot which, at its peak, employed 4000 people. Its decline followed the 1982 Defence Review when its future was already in doubt due to increasingly stringent explosives regulations and more complex missile technologies. The site was closed in 1989.

   The site has three elements. The 22ha heritage area on the site of the original conserved depot now houses Explosion, a naval armaments museum. The heritage area has both EC and Millennium funding. The car park is the northern limit of the trail.

   The second, socially more significant element, is the development area on the former safety zone around the depot. This area is where three major companies are constructing 700 dwellings. House prices are cheaper in Gosport and therefore the social mix will contrast markedly with that at Gunwharf. The names such as Admirals Quay, Mariners Wharf, Crown Point, The Spinnakers, Vanguard Way and Challenger Road all conjure up images of maritime and military power.

   The third, smaller area, adjacent to the Harbour and Forton Lake is being left as open space because it abuts the SSSI, Ramsar site and SPA where waders and wildfowl spend the winter. Retrace your route back to the ferry terminal via Mumby Road and Falkland Gardens.

Bibliography and further information

© Portsmouth City Council

Ashworth, G.J. (1991) *War and the City.* London: Routledge.

Bateman, M. and Riley R.C. (eds) (1987) *Geography of Defence.* London: Routledge.

Burtenshaw, D., Bateman, M. and Ashworth, G.J. (1991) *The European City: A western perspective.* London: David Fulton.

Coad, J. (1993) *Deserted Bastions: Historic naval and military architecture.* London: SAVE.

Coad, J. (1998) 'The remarkable built heritage of Naval dockyards' in *Conference Proceedings RENDOC 1998.* Chatham: Chatham Historic Dockyard Trust, pp. 6-15.

Portsmouth City Council (1991) *1991 Census Ward Profiles.* Portsmouth: Portsmouth City Council.

Portsmouth City Council (1991) *An Everyday Atlas of Portsmouth.* Portsmouth: Portsmouth City Council.

Portsmouth City Council (2000) *Portsmouth City Local Plan, First Review 2001-2011, First Deposit Draft.* Portsmouth: Portsmouth City Council.

Portsmouth City Council *et al.* (Commissioners) (2000) *Portsmouth Harbour Plan Review 2000.* Portsmouth: Portsmouth City Council.

Riley, R.C. (1972) *The growth of Southsea as a naval satellite and Victorian resort. Portsmouth Papers 16.* Portsmouth: Portsmouth City Council.

Riley, R.C. (1985) *The evolution of the docks and industrial buildings in Portsmouth Royal Dockyard 1698-1914. Portsmouth Papers 44.* Portsmouth: Portsmouth City Council.

Riley, R.C. (1987) 'Urban conservation or private museum? Historic architecture in Portsmouth Dockyard', *Urban Conservation: International Comparisons.* Occasional Paper, 7, pp. 8-13.

Riley, R.C. (ed) (1987) *Portsmouth Field Guide.* Portsmouth: Portsmouth Polytechnic.

Stapleton, B. and Thomas J.H. (eds) (1989) *The Portsmouth Region.* Gloucester: Alan Sutton.

Tourist Information Centres

Central Library
Guildhall Square
Portsmouth PO1 2AL
Tel: 023 9283 8382

The Hard
Portsmouth PO1 3QJ
Tel: 023 9282 6722

Clarence Esplanade
Southsea
Portsmouth PO5 3PB
Tel: 023 9283 2464

1 High Street
Gosport PO12 1BX
Tel: 023 9252 2944

Maps

Geographer's A-Z Map Company *A-Z Street Atlas Portsmouth* (1:15,840)
OS Landranger 196 The Solent & Isle of Wight (1:50,000)
OS Landranger 197 Chichester and the Downs (1:50,000)

Websites

Dockyard
www.flagship.org.uk

English Heritage
www.english-heritage.org.uk

Gosport Borough Council
www.gosport.gov.uk

Gunwharf Quays
www.gunwharf-quays.com

Hampshire County Council
www.hants.gov.uk

Institute of Maritime and Heritage Studies
www.envf.port.ac.uk/maritimeheritage

Portsmouth&
www.portsmouthand.co.uk

Portsmouth City Council
www.portsmouth.gov.uk

Portsmouth Harbour Project
www.envf.port.ac.uk/geo/research/portsmouth/port1.htm

Portsmouth News
www.thenews.co.uk

Portsmouth Society
www.portsmouthsociety.org.uk

Southsea Town Council
www.southseatowncouncil.co.uk

Local Studies libraries

Local History and Naval Collection
Central Library
Guildhall Square
Portsmouth PO1 2DX
Tel: 023 9281 9311

Frewen Library
University of Portsmouth
Cambridge Road
Portsmouth PO1 2ST
Tel: 023 9284 3222

Photo credits

The GA would like to thank the following for their help with photographs for this book:

British Library

Camper & Nicholsons Marinas Ltd

David Burtenshaw

English Heritage Photo Library

Gosport Borough Council

Portsmouth City Council

Portsmouth City Museum and Records Office

QinetiQ

Royal Armouries

RN Submarine Museum, Gosport

Taylor Woodrow Property Company Ltd

University of Portsmouth Printing and Photographic Services